I0750692

Paper Book of the Defendants.

COMMONWEALTH *vs.* THE PRESIDENT, MANAGERS AND COMPANY, for erecting a Bridge over the Allegheny River, opposite the City of Pittsburgh, in the County of Allegheny.	In the Supreme Court of Pennsylvania, for the Western District of the Term of September and October, 1850. Information of JAMES TODD, praying a Writ of Quo Warranto.

The application for a Writ of Quo Warranto, in this case, is addressed to the sound discretion of the Court.

The Writ, though usually sought in the Court of Common Pleas, *may* be issued out of the Supreme Court, "with the leave of the said Court, in term time, or of any Judge of said Court in vacation." The Court are not bound to issue the Writ as a Writ of right, either at the instance of the Attorney General, his Deputy, or a private prosecutor. There must be at least a clear prima facie forfeiture of corporate rights, privileges or franchises, by misuser or nonuser, shewn upon the face of the information, to justify the allowance of the Writ. The acts or omissions producing the forfeiture should be fully and distinctly stated. Their sufficiency to work a forfeiture and justify a judgment of seizure of corporate franchises, ought to be manifest and palpable, conceding entire verity to the allegations contained in the information. The defendants deny their sufficiency, uncontradicted and unexplaned, to justify a forfeiture of their corporate franchises. They show no neglect of duty or violation of right requiring the interposition of the Court. That interposition is not invoked by the Attorney General, Auditor-General, or any other accredited functionary of the Commonwealth. It will be shewn hereafter that *they* seek no redress for violation of public rights. After examination they are satisfied that the defendants have neither done or omitted any acts injuriously affecting the public interest, or demanding their official action.

The prayer for the Writ emanates from a person holding no official station, but desiring, as he may legally do, to prosecute the same. If any wrong be charged in the information, it is a private wrong. If there be any remedy, it is by individual suit or action. It is sought by this proceeding to jeopard and forfeit rights and privileges, the acquisition of which involved an expenditure of more than one hundred thousand dollars; to divest helpless widows of their property, and deprive tender orphans of their support; to divest and destroy the gushing and refreshing streams which God-like Charity has destined for the relief and sustenance of helpless sorrow and suffering, because the relator has been compelled to pay, in tolls, what the original and amended Charter expressly authorized. Connecting the Charter with the information, as prayed by the relator, the first charge, however amplified in allegation or involved in expression, amounts but simply to a charge against the relator of the legal rate of toll, without the distinct imputation that the relator suffered pecuniary loss or personal injury, and without any intimation that the public were prejudiced or affected thereby.

We now proceed to a brief examination of the rights, powers and duties of the Company, under its Charter.

The second section provides that upon the happening of certain events, the Governor shall constitute the subscribers, a body corporate or politic by the name and style of "The President, Managers and Company, for erecting a Bridge over the Allegheny River, opposite Pittsburgh, in the county of Allegheny," with all the privileges incident to a corporation, who shall have perpetual succession, and shall be capable of taking and holding their said capital stock, and the income and profits thereof, and of enlarging the same by new subscriptions, if such enlargement shall be necessary to fulfil the purposes of the act. This provision secured to the Company the right of property in their stock, and placed its income and profits under the protection of law. Its rights might be declared forfeit and lost by certain acts and omissions, known to the law as causes of forfeiture, and defined and recognized as such by a regular course of judicial proceeding. But such cause of forfeiture would never be presumed without full and ample proof. The party to be prejudiced would never be called upon to answer without clear, distinct and positive allegation of adequate causes of forfeiture. The third section gives the Company power to elect officers to conduct the business of the Company, and to make such bye-laws, rules, orders and regulations (not inconsistent with the Constitution and laws of this State or United States), as may be necessary for the well ordering of the affairs of this Company. The Company is, by this provision, made the judge of what bye-laws, rules, orders and regulations may be necessary for the well ordering of its affairs. None such will be declared causes of forfeiture, when sustained by the letter or spirit of the Charter, or in harmony with its provisions. None such will be declared causes of forfeiture at Common Law, unless clearly shewn to be such by legal principles and authoritative adjudications distinctly applicable.

The sixth section gives to the President and Managers "power and authority to appoint such engineers, superintendants, assistants, and workmen as they shall deem necessary to the erection of such Bridge." The eighth to take materials for the erection of the bridge.

The tenth section provides that when a good and complete bridge, under the authority of this act, shall be erected over the Allegheny River at Pittsburgh, the property of the same shall be vested in the said Incorporated Company, their successors and assigns, and the said Company, their successors and assigns are hereby empowered to erect gates, demand and receive tolls as follows, viz: for *every foot passenger two cents, &c.*

The eleventh section imposes a penalty of twenty dollars for taking excessive toll and neglecting to keep the bridge in repair. The fifteenth section requires the bridge to be completed within seven years from the approval of the act, March 20th, 1810, or if the Company fail to do, that gives the Legislature power to resume all and singular the rights and privileges thereby granted to the Company.

Under this Charter the requisite amount of stock could not be obtained —desirable as was the proposed improvement to the public; individuals were not willing to hazard their funds in the adventure. The project languished and would have ultimately failed had not the Legislature interposed by legislative enactments and pecuniary assistance, and public countenance and credit.

Accordingly on the 16th February, 1816, a new act was passed. The first section recites the expiration of the original Charter by its own limitation, re-enacts its provisions and declares it perpetual. The second provides for a subscription of sixteen hundred shares by the Commonwealth to the stock of the Company, "the one-half to be paid when the piers and abutments of the bridge should be constructed, and the other half when the superstructure should be raised. This subscription secured to the Company, from a reliable source, forty thousand dollars. It gave elasticity, vigor and availability to the efforts of individuals, and ultimate success to the long protracted enterprise.

Forty-four thousand dollars of stock were subscribed by individuals constituting, with that subscribed by the Governor on behalf of the Commonwealth, an aggregate of eighty-four thousand dollars, upon the strength, faith and credit of which the work was commenced, conducted and completed. Its construction resulted in an expense of one hundred and one thousand dollars, exceeding by seventeen thousand dollars the stock owned by the Company, and in the contracting of a debt by the Company to the latter amount, which required liquidation before any dividends or profits could be realized either by the Commonwealth or individuals.

The seventh section empowered the Company to erect gates, and demand and receive tolls as follows, to wit: for every foot passenger two cents, &c., and exempted from the payment of toll, any person attending funerals, any detachment of the military of this State, or the United States, foot passengers attending divine service on the Sabbath Day, students or children attending schools or other seminaries of learning." The ninth section extended the term of finishing the bridge ten years, &c.

The great object of these statutory provisions was to secure to the public the convenience of an easy, ready and uninterrupted passage across the Allegheny River, safe and comfortable at all times and under all circumstances, for a compensation fair and not exceeding a specified amount. That object was attained. The bridge was erected. The relator does not charge that the bridge was not erected within the time required by the Charter. He does not complain that its quality, character and convenience were not such as the public convenience and expectation desired, or as the terms of the Charter exacted. The public fully participated in the enjoyment of all that was anticipated from the erection and completion of the bridge. No complaint is made that, from the day of its completion to the present, the bridge has not been kept in perfect repair. No allegation is made that it has not at all times and under all circumstances, afforded to all people who desired, a safe and easy transit for their persons and property.

It is not charged that any one has been prevented from passing the bridge, upon payment or tender of the stipulated compensation, or hindered or delayed, upon any pretext, from so doing. The Charter expressly authorizes the Company to charge for *every foot passenger two cents.*

This is an express statutory right clear, distinct, unambiguous and unequivocal.

It is universal (excepting the cases mentioned in the proviso of the seventh section of the act of the 17th of February, 1816), in its application. It reached the relator as well as all other individuals. The relator does not charge that he was ever prevented from passing the bridge. He does not complain that more was exacted from him than he was bound to pay; that any excessive toll was ever claimed from him. He does not charge that the Company have not awarded to him all his rights under the Charter. If any person passes the bridge for a less sum than that specified in the Charter, or without payment at the time of passage, it is of favor and not of right.

The Company are not bound to commute. They may insist upon prompt payment from passengers. But convenience has dictated the policy of such arrangement; but they were not bound, if desired, to accord it to the relator. He was, however, at one time, permitted to avail himself of the arrangement. Had he behaved as others did, the arrangement would, if desired, have been continued with him. His name was excluded for cause. To prevent disturbance and difficulty, the Company insisted upon prompt payment from him, and enforced their legal rights. The relator then charges no violation of private right, or pecuniary injury. He does indeed charge, and this appears to be the entire gravamen of his complaint, that said *invidious distinctions* operate as a *virtual* prohibition and exclusion of the relator from a fair and just competition in the course of honorable industry with others of his occupation, which is that of a Carpenter. He does not charge that he pays in the aggregate more than three dollars per annum. He does not aver that he loses a single penny by the exclusion. He does not allege that the Allegheny river runs between his residence and the place of his employment. He has not informed us where he resides, where he is employed, whether he necessarily passes the bridge in the course of his employment, whether, if necessarily passing the river, it is more or less convenient to pass the Allegheny than the Hand street Bridge. It is, in truth, not shown that it is essential or important to fair competition in the carpenter business to pass the bridge at all. For aught that appears, that business may be as successfully and profitably followed on that side of the river on which a man may reside, as on the opposite; and yet from these loose deductions, these vague inferences, these unsupported conclusions, the relator avers a *misuser* of the rights, privileges and franchises conferred by the Acts of Assembly, a violation of the trust confided by the Commonwealth, and a forfeiture of the corporate franchise. For this imputed insult to the sovereignty o the relator, the majesty of the law is invoked to deprive the stockholders of their corporate rights, and to deprive the public of the right and convenience of transit over a structure, the want of which was long and severely felt, and the use of which has contributed immensely to the public improvement and prosperity, and the loss of which would be keenly felt by the countless thousands who now enjoy its benefits. The Court is emphatically asked to forfeit private rights reaching one hundred thousand dollars in value, and to sacrifice public advantages of inestimable and inappreciable importance, for an alleged, inferential private injury to the relator. The injury, if any, to the relator is strictly private, and for such injury the Court will not forfeit a corporate franchise.

The proceeding is not sustained by any authority quoted. The case of The People *vs.* The Hillsdale and Chatham Turnpike Company, 2 *Johnson's Reports,* 190, is in our judgment a conclusive authority against it. "WOODWORTH, Attorney-General, moved for a rule that the defendants show cause, by the next term, why an information, in the nature of a *Quo Warranto,* should not be filed against them. He read affidavits stating that the road had been opened through the land of the complainants, and used, without any offer having been made to them to agree upon the compensation, and without having the damages ascertained according to law.

"E. WILLIAMS, contra.

"Per CURIAM. If the defendants have not followed the directions of the Act relative to the compensation to be made to the owners of the land, through which the road had been made, they are trespassers, and the complainants have adequate remedy in the usual course of the common law. The public are no way interested in the controversy or complaint, and that is a sufficient reason for not granting this extraordinary remedy.

"Rule refused."

In this case it will be noticed the Attorney-General moved for the rule. The affidavits showed a violation of law by the corporation. A palpable injury had been done by the company to an individual. But because the *complainants had remedy by due course of law, and because the public was no way interested in the controversy or complaint,* the rule was refused. The injury in that case was private only. The remedy was consequently by private action. So here, the injury, if any, is strictly private. The rights of the public have not been violated, and consequently a private individual cannot successfully or properly employ the name or invoke the aid of the Commonwealth.

We concede that a private and a public remedy may co-exist. A private action and a public prosecution may in some cases both be sustained.

The same doctrine is laid down in the case of The People *vs.* Bristol and Renssalaerville Turnpike Company, 23 *Wendell's Reports*, 244–5. The principle now contended for (that there might be a remedy both by private action and public prosecution,) was argued in London *vs.* Vanacre, 12 *Mod.* 270–1. It was held, as we before noticed, that the city must forfeit its franchise of the shrieverick if it did not elect a sheriff and compel him to serve. He had refused to serve. HOLT said as to the objection that he may be indicted for this refusal, as was the case of Lanwood, sheriff of Norwich: I answer, that will not be sufficient to hinder the forfeiture of the franchise; for if there should be a vacancy when the sheriff comes to be sworn, there will be an obstruction of justice. "*The ground was that a public inconvenience in not exercising the franchise would work a forfeiture, although there might be another remedy.*"

These cases clearly show the instances in which a private action and a public prosecution may both be sustained. This remedy is sought in the name and under the authority of the Commonwealth, and cannot, upon the strength or reasoning of any authority adduced, be sustained upon allegation of any mere private injury.

We concede that "all franchises which are granted, are upon condition they shall be duly executed *according to the Charter, that being a condition annexed to the grant.*" This Company is not charged with the doing of any act prohibited by its Charter; with any infraction of the law of its being.

"*The Charter is the law of the case.* If none of *its* provisions have been violated, it is difficult to find any legitimate ground to demand its surrender."—*Corwin versus Urbana Insurance Company*, 14 *Ohio Reports*, 10.

No case has been adduced, none it is believed can be found, where the franchise of a Company for the construction of a road or bridge have been seized or forfeited for *misuser*, where the act or acts were not specially and distinctly charged, and when the act or acts did not injuriously affect public rights and interests. The law of New York provides that if corporations offend against "any of the provisions of the act or acts creating" them, the information may be filed and judgment of ouster rendered.—23 *Wendell Rep.*, 208.

The authorities quoted from Story probably exhibit very sound equity, but we deny their applicability to this controversy. The information charges the adoption of a resolution by the corporation prohibiting the enrolment of the relator's name.

The Charter expressly authorizes the Company to make such bye-laws, rules, orders and regulations (not contrary to the law, &c.,) as may be necessary for the well ordering of the affairs of the Company.

The information does not charge that the resolution excluding the relator, was a violation of this provision. It will not be presumed such until the contrary is alleged and shewn, especially in the absence of any imputation of exceptionable motives.

There is nothing in the Charter prohibiting the resolution; there is nothing in the information shewing that it was adopted through the influence of any reprehensible design, or without abundant and justifi able cause. Conceding the correctness of all the authorities quoted by our opponents, they do not as we conceive, sustain the positions which they desire to establish. Most of them are wholly irrevelent. Never was a corporation put upon its defence under analagous circumstances, at least never any that has come within our reading or observation. One single remark more, and we dismiss the consideration of the charge of misuser. The public have a deep and vital interest in this proceeding. If the rights of the corporation are forfeited, and its franchises seized, the right of taking toll, &c., and all other corporate action is gone, individual interests will be seriously affected. The right of the public, resulting from the enjoyment and exercise of the corporate rights and privileges of the Company, to transit and passage over the bridge, will be extinguished. A grievous public calamity would inevitably result from the redress, sought for an alleged private injury. In the fifth proposition laid down by the relator's counsel, it is said to be "a principle of sound *law*, as well as public policy, that *next* to courts doing right in the exercise of their high and sacred functions, they should give public satisfaction. A proceeding that should deprive individuals of their property, and the community of their right and convenience of transit over the bridge, would give little public satisfaction, and secure very few blessings, upon the tribunal which should achieve so disastrous a result. To sacrifice one hundred thousand dollars of property, to deprive our citizens of a great public convenience, to exclude from the State Treasury seven hundred dollars per annum, now received from the Company in taxes, because the relator saw proper to exercise his sovereignty in a fight with the toll gatherer, would indeed be a measure of redress deserving little commendation for its intrinsic justice or practical influence.

The facts alleged in the second branch of the relator's information are not charged as constituting a *misuser* or *nonuser*, and working a forfeiture of the corporate franchises, but are mere allegations that in certain events and contingencies, which the Company were not bound regard, the bridge might have been free. This, however, is mere conjecture.

Were it true, it would afford no valid reason for the allowance of the proposed Writ. This Court cannot declare the bridge free in the proposed proceeding, nor adjudge its franchises forfeit upon the allegation and proof of facts that it might and ought to be made free. A forfeiture of corporate rights and privileges is a very different thing from the accumulation of means adequate to a full and just compensation of stockholders, and justifying the dedication of the structure to public use.

These two things are essentially different in their nature, quality and legal characteristics. The one rests upon allegation, proof and judicial determination ; the other requires not only allegation and proof, but also legislative enactment. The one may be accomplished through the instrumentality of judicial proceedings ; the other cannot be perfected, even if it can be legally commenced without legislative action. It will be shown hereafter that legislation has been procured with reference to the question of making the bridge free, and ample means provided for a thorough investigation through the agency and under the supervision of the District Court.

It will, however, be apparent, upon a comparison of the latter branch of the information with the provisions of the Charter, that the Company are not in the supposed default. By the twelfth section of the Charter it is provided, that the Company shall " declare and make a dividend of the income and profits thereof, among all the subscribers to the said Company's stock, in proportion to their respective shares, first deducting all contingent costs and charges, and such proportion of the said income as may be sufficient for a fund, to provide against the decay, the repairing or rebuilding of the said bridge, as time and accident may render necessary." This provision expressly authorizes a deduction for costs and charges, and of an amount sufficient to repair and rebuild. It is not charged or pretended that the accumulations undivided were sufficient to accomplish these purposes, or to render the bridge free. In truth, they are not now, and never were, sufficient. But it is alleged, and this appears the burthen of the charge, that the excess of fifteen per cent. dividends which the corporation *might have received* if they had not violated their trust in remitting, without authority of law, the tolls imposed by the Act aforesaid, would have been more than sufficient to redeem the bridge. This position assumes that the Company have no power to reduce or remit tolls; no authority to commute or arrange the amount—the whole amount allowed by the Charter must, under all circumstances, be exacted. Neither mercy, kindness or charity is to be tolerated. Hoary age, helpless infancy, abject poverty, each, all must pay the full amount that might be exacted by the Charter. Under no circumstances can there be deduction, discrimination or dispensation. Though the profits of the stockholders would be augmented, the interests of the public promoted, the period of redemption accelerated thereby, the Directors are powerless, and cannot commute or reduce the tolls from the highest limits fixed by the Charter. The kind hearted clergyman, passing to administer the last rites and consolations of religion; devoted woman, hastening, from the very impulse of her nature, to the relief of sorrow and suffering, must, in their holy missions, become the objects and victims of indiscriminate exaction. The doctrine assumed in this branch of the information is utterly subversive of the position upon which the relator rested in the first. In that it was claimed that the franchises ought to be forfeit, because the relator was not permitted to participate in the benefits of commutation; in this the allowance of the Writ is urged, because the full amount of toll allowed by the Charter has not been collected from every one.

It is admitted that there have been judicious reductions of tolls by the Company, and in every instance their receipts have been thereby increased. Competition has rendered such reduction necessary. Correct policy would fully justify it. Without reduction the business and receipts would have rapidly diminished, and the bridge would never be free. Discriminations have been made, and are avowed and tolerated. Clergyman and females are permitted to pass free of charge. And under this judicious system a fund, carefully invested in Bank and Gas Stock, approaching thirty thousand dollars has been accumulated.

The redemption of the bridge, it is confidently believed, would not be accelerated by any interference with the present judicious arrangements and careful management of the Company.

There surely is nothing in the Charter which requires the Directors to charge and receive the rate of tolls specified. It is permissive, not compulsory. More may not be exacted; less is not prohibited. Extortion and oppression are prohibited under penalties. Danger might well be apprehended from charging too much, never from charging too little. The discretion exercised by the stockholders is, we think, fully justified by the letter, spirit and design of the Charter.

The fourteenth section provides that the "Company shall be oblige to take such sum of money therefor (the bridge) as shall be allowed on a fair appraisement by disinterested persons, to be appointed in such manner as shall *be directed by law*." Legislation is required previous to final action. We now proceed to show that ample provision has been made, by legislative enactment, for a thorough investigation of the affairs of the Company, and a determination of all issues of law and fact that may arise, in the District Court of Allegheny county. On pages 485–6, Pamphlet Laws of 1846, Secs. 5, 6, and 7, the following provisions will be found :

"SEC. 5. The District Court of Allegheny county is hereby vested with full authority and jurisdiction to examine into and determine all issues of law, or fact, which may in any way arise in the investigation of the affairs of the Company chartered by the Legislature on the 20th day of March, A. D. one thousand eight hundred and ten, to erect a bridge over the Allegheny river, opposite the city of Pittsburgh, in order to determine if said bridge, according to the provisions and conditions of said Charter, should not now be a free bridge."

"SEC. 6. That said Court is hereby authorized, if deemed expedient, to appoint three disinterested and competent men to examine into the receipts and expenditures of said bridge Company, since its construction ; the said commissioners being first duly sworn or affirmed, shall have power to examine before them, under oath, any person or persons touching the affairs of said Company ; to call for and enforce by attachment, or otherwise, the production of all books and papers of said corporation which may be useful, in order to ascertain the cost of the bridge, the amount of the annual receipts and expenditures of said Company, the amount of dividends, and the rate per cent. per annum paid to the stockholders, the amount invested as a contingent fund, together with such additional information as may be pertinent to the issues involved; which report, if confirmed by said Court, shall be deemed and taken to be final and conclusive, in any issue which may hereafter arise in regard to the franchise of said corporation."

"SEC. 7. Said commissioners shall each be entitled to receive one dollar and fifty cents per day for every day by them necessarily occupied in said investigation, to be paid as may be directed by said Court, from the treasury of Allegheny county, or from the funds of said corporation."

It might well be contended that this remedy, so full, appropriate and complete, was designed by the Legislature as a substitute for all others, or at least as necessarily preceding action by any other. It is certainly an exceedingly appropriate provision for ascertaining the true condition of the Company, with a view to the redemption of the bridge. It gives full authority and jurisdiction to the District Court of Allegheny county to examine into and determine all issues of law or fact which may in any way arise in the investigation of the affairs of the Company. If this remedy be not exclusive it is at least concurrent, and far more convenient and appropriate than a proceeding in the Supreme Court. We believe that this jurisdiction was conferred as an appropriate, convenient and exclusive remedy; but if the Court should be of a different opinion it is believed that the relator would be remitted, at once, to a tribunal so convenient and so entirely competent, by the aid of the facilities afforded of giving early and complete redress. Such was the view of the Court of Common Pleas, after a full, patient and thorough investigation, as expressed in an exceedingly able, elaborate and well-considered opinion. The countenance of the Attorney General has been invoked to the prosecution of charges against this Company. A correspondence between him and the Auditor General has taken place in reference thereto. To dispel misconceptions and misunderstandings, to counteract misrepresentation and misapprehensions, to exhibit clearly and correctly the conduct and condition of the Company, that correspondence will be presented to the Court.

The Attorney General having communicated to the Auditor General the letter of Messrs. Woods and Alden, the Auditor General wrote to the Attorney General as follows:

Auditor General's Office, *Harrisburg*, 24th June, 1850.

Dear Sir,—Your letter relating to the claim of the Commonwealth against the Allegheny Bridge Company, (the old bridge at St. Clair street,) has been received. In reply, have to say that at present I have not time to make a thorough investigation of the claim, but shall do so at my first leisure, perhaps during this week. If it is ascertained that the claim is well founded, it would be proper for the accountant officer to settle an account charging the Company with whatever amount may be due. Authority for this course will be found in the Act of 1811, relating to the settlement of public accounts, and also in the case of Commonwealth *vs.* Easton Bank. You will please consider yourself professionally concerned for the Commonwealth. And I shall feel obliged to you for any information which may conduce to a speedy and just adjustment of the claim.

I am very respectfully,

JOHN N. PURVIANCE, *Aud. Gen.*

Hon. C. Darragh, *Atty. Gen.*

On the 31st of July the Auditor General wrote as follows:

Auditor General's Office, *Harrisburg*, 31st July, 1850.

Dear Sir,—In my letter to you of 24th of June last, I promised I would write again in reference to the supposed claim of the Commonwealth against the Old Allegheny Bridge Company.

Upon examination of the proper papers in this Department, it appears the Company have made, regularly, returns of their dividends, and paid the tax to the Commonwealth upon the capital stock proportionate to said dividends, as directed by law. Unless, then, the returns are inaccurate, copies of which, if you think it essential, I would send you, there is nothing due from the Company to the Commonwealth.

I am, with great respect, yours, &c.,

JOHN N. PURVIANCE, *Aud. Gen.*

Hon. C. Darragh, *Atty. Gen.*

It is probable if there is any claim that it is for the Commonwealth's portion of dividends prior to the sale of the stock owned by the Commonwealth.

After this one of the counsel of the Company proposed to the Attorney General that the Secretary of the Bridge Company should wait upon him with the records and papers of the Company; that the Attorney General should examine them, and after such examination communicate the result to the Auditor General. He acceded to this proposition, the Secretary waited upon him with the books and papers, and after examination he wrote to the Auditor General the letter which follows, the counsel of the Company not being present at the time of examination, or having had any communication with him after his acceding to the aforesaid proposition. He kindly permitted a copy to be taken of his letter to the Auditor General, which is as follows:

Pittsburgh, 23rd August, 1850.

John N. Purviance, Esq., Auditor General,

Sir,—Some time ago I sent you a communication, signed by T. J. Fox Alden and Robt. Woods, Esqs., having reference to a supposed interest of the State in the affairs and funds of the Allegheny Bridge Company At the request of A. W. Loomis, Esq., the counsel for the Bridge Company, I have made an examination, as far as I could, of the books, *minutes* and statements of the Company, since the date of the act of incorporation to the present time, and I am satisfied justice has been done the State by the Corporation. For the last five or six years the maximum dividends allowed by the Charter (15 per cent.) has been paid to the stockholders, but it is quite clear that the dividends average from the date of the Charter to the present time, does not exceed seven per cent. on the cost of the bridge.

it appears from the books of the Company that the dividends coming to the State were regularly paid, whenever a dividend was declared; and it will be observed that no dividend was declared from 1815 to 1824. Some few years since the State sold out *their* stock at a large advance on the original price. The State paid $25 a share, and sold for $33; and the purchasers of the State stock at this advanced price, believe that they ought to be permitted to receive such dividends as the Charter and the affairs of the Company allow. The Corporation has made regular returns of their affairs to the office of the Auditor General. When the State sold out *their* stock, Mr. Packer, then Auditor General, wrote to the Company that as the State had no further interest in the bridge, the returns could be dispensed with; but the Company thought it better to continue the annual returns, and you will find them in your office.

I write this letter at the request of the persons interested in the bridge, and to express to you the opinion, that so far as I have been able to see, the State has no reason to complain of the management of the affairs of the Bridge Company.

Very truly yours,

C. DARRAGH.

The letter of the Auditor General Packer, referred to in the letter of the Attorney General, is as follows:

AUDITOR-GENERAL'S OFFICE, *Harrisburg*, Dec. 20, 1844.

Dear Sir,—Since the State has disposed of her stock in the Allegheny Bridge Company, a statement of the receipts and expenditures of that Company will not be required, in pursuance of the Act of 27th March, 1824.

If your act of incorporation, therefore, does not direct it, you may hereafter dispense with forwarding such report. I have not your act of incorporation before me.

Yours respectfully,

W. F. PACKER, Audr. Genl.

J. HARPER, Esq.

But the Company, apprehensive that difficulties might arise if returns should be omitted, adopted the following proceedings:

The following preamble and resolution is taken from the minutes of the Board of Managers of the Allegheny Bridge, of April 6th, 1846:

"Whereas Wm. F. Packer, Esq., Auditor General of the State of Pennsylvania, did on the 20th day of December, 1844, address a letter to the Treasurer of this Company, instructing him to discontinue the annual reports of receipts and expenditures, as being unnecessary since the sale of the State stock; and whereas, the act of 27th of March, 1824, remains still in force, and it may be doubtful how far it may operate as concerns the bridge, therefore,

"Resolved: That the Treasurer be required to continue his annual statements to the Auditor General, or other proper officer, and that he request an acknowledgment of the receipt of his report for 1845."

The Treasurer made the following statement when a similar application was before the Court of Common Pleas:

NOTE BY THE TREASURER.

I find among the papers of the Company, copies of the reports forwarded to the Auditor General, by my predecessor, John Snyder, Esq., and since I became Treasurer in 1842, I have forwarded annual reports to Harrisburgh, of the affairs of the Company for the year, which have always been sworn to before an Alderman.

J. HARPER, Treas. A. B. Co.

Jan'y. 20, 1850.

It was in consequence of this prudent and judicious resolution of the Company, that the returns referred to by the Attorney-General were maee. On the hearing before the Common Pleas, a full statement of receipts, dividends, and other affairs of the Company was made, portions of which have appeared in the public papers. There is no desire to suppress or conceal. The Company are ready and willing to submit to fair examination their conduct and condition. They have ever been ready and willing, and now are, to meet the investigation contemplated by the act of 1846. With the utmost cheerfulness, every facility for a full and thorough examination, will be afforded.

The foregoing objections to the allowance of the Writ prayed for by the relator, are to our minds satisfactory. We hope and trust they will be regarded as sufficient by the Court.

We have also desired to vindicate the Company from undeserved reproach and heedless and vindictive aspersion. We have desired to exhibit our clients as they are, pursuing the even tenor of their way in the onward course of truth, honor and justice, discharging their duties to the Commonwealth, their constituents, the public and themselves faithfully, conscientiously and fearlessly. They ask no favor; they deserve no reproach.

They have been, and have been *justly*, esteemed and regarded as public benefactors. Once the whisper of suspicion the voice of reproach was not heard. At one time they would not have been tolerated.

When from 1810 to 1816 they struggled to secure the funds requisite to the success of a most important public improvement; when from 1816 to to 1819 they contributed freely and liberally the means which secured the construction and completion of a work which has yielded countless comforts and conveniences to the citizens of Allegheny and Pittsburgh, and contributed in an eminent degree to the mighty improvements which have so essentially promoted the public interests; whilst from 1819 to 1824 they beheld *others* enjoying the conveniences and comforts resulting from their efforts and expenditures, without *profit or compensation* to themselves, *then*, indeed, they were regarded as worthy men and public benefactors. They were then deemed worthy to live, move and have a being among their fellow citizens. Then their conduct and motives were spared from imputation and reproach. But when in the course of time, after the lapse of many years, they began themselves to realize liberal profits from their efforts and investments, imputation and reproach are abundant. Envy, hatred, malice, and all uncharitableness, become prevalent. Purity of motive, propriety of conduct, will not be conceded by that spirit of envy and malignity that discovers no merit in any motives, purposes or actions but its own. This Company has already paid an abundant tribute to the unfriendly designs and purposes of covert prejudice and open hostility. It will not shun fair and legitimate investigation. It desires to pursue in peace and quietness its onward course, yielding to constituents the legitimate fruits of their investments, discharging all obligations to the public, and accumulating, as rapidly as may be compatible with existing rights and duties, the means of ultimate redemption of that structure which was long the object of ceaseless effort and lavish expenditure, which ever has been, and which, it is hoped and believed, will, through coming years, continue to be the source and means of multiplied and increasing blessings.

That spirit which would defeat and deny its peaceful and lawful purposes, which would impair or destroy its rights and franchises, which would clench the hand when relaxing for the relief of indigence and misfortune, which would harden the heart when yielding to the appeals of charity and kindness—that spirit, it is hoped and believed, will be emphatically rebuked and effectually restrained.

METCALF & LOOMIS,
DEFENDANTS' ATTORNEYS.

www.ingramcontent.com/pod-product-compliance
Lightning Source LLC
LaVergne TN
LVHW020637110826
845149LV00004B/1247

* 9 7 8 1 4 1 8 1 9 1 2 1 4 *